BUBBLE·OLOGY

Teacher's Guide

Grades 5–8

Skills
Observing, Measuring and Recording Data, Experimenting,
Classifying, Drawing Conclusions, Controlling Variables,
Calculating Averages, Graphing Results

Concepts
Technology, Engineering, Chemical Composition, Substances, Properties,
Surface Tension, Hygroscopicity, Optimum Amount, Bernoulli's Principle,
Aerodynamics, Pressure, Patterns, Light and Color, Interference,
Air Currents, Evaporation, Environments

Themes
Systems & Interactions, Stability, Patterns of Change,
Scale, Structure, Matter

Mathematics Strands
Measurement, Geometry, Statistics and Probability,
Patterns and Functions

Nature of Science and Mathematics
Scientific Community, Interdisciplinary, Cooperative Efforts,
Creativity & Constraints, Theory-Based and Testable,
Changing Nature of Facts and Theories,
Objectivity & Ethics, Real-Life Application, Science and Technology

Time
From eight to ten 45- to 60-minute sessions

Jacqueline Barber

LHS GEMS

Great Explorations in Math and Science (GEMS)
Lawrence Hall of Science
University of California at Berkeley

Illustrations
Lisa Haderlie Baker
Carol Bevilacqua
Janice A. Coonrod
Wendy Kitamata-Pulaski
Lisa Klofkorn

Photographs
Jacqueline Barber
Cary I. Sneider

Lawrence Hall of Science, University of California,
Berkeley, CA 94720. Chairman: Glenn T. Seaborg;
Director: Marian C. Diamond

Initial support for the origination and publication of the
GEMS series was provided by the A.W. Mellon
Foundation and the Carnegie Corporation of New York.
GEMS has also received support from the McDonnell-
Douglas Foundation and the McDonnell-Douglas
Employees Community Fund, the Hewlett Packard
Company Foundation, and the people at Chevron USA.
GEMS also gratefully acknowledges the contribution of
word processing equipment from Apple Computer, Inc.
This support does not imply responsibility for statements
or views expressed in publications of the GEMS program.
Under a grant from the National Science Foundation,
GEMS Leader's Workshops have been held across the
country. For further information on GEMS leadership
opportunities, or to receive a publication brochure and the
GEMS Network News, please contact GEMS at the address
and phone number below.

International Standard Book Number: 0-924886-58-7

You can also reach GEMS by e-mail at
gems@uclink.berkeley.edu
or visit our web site at www.lhsgems.org

COMMENTS WELCOME

Great Explorations in Math and Science (GEMS) is an
ongoing curriculum development project. GEMS
guides are revised periodically, to incorporate teacher
comments and new approaches. We welcome your
criticisms, suggestions, helpful hints, and any
anecdotes about your experience presenting GEMS
activities. Your suggestions will be reviewed each
time a GEMS guide is revised. Please send your
comments to: GEMS Revisions, c/o Lawrence Hall of
Science, University of California, Berkeley, CA 94720.
The phone number is (510) 642-7771.

Great Explorations in Math and Science (GEMS) Program

The Lawrence Hall of Science (LHS) is a public science center on the University of California at Berkeley campus. LHS offers a full program of activities for the public, including workshops and classes, exhibits, films, lectures, and special events. LHS is also a center for teacher education and curriculum research and development.

Over the years, LHS staff have developed a multitude of activities, assembly programs, classes, and interactive exhibits. These programs have proven to be successful at the Hall and should be useful to schools, other science centers, museums, and community groups. A number of these guided-discovery activities have been published under the Great Explorations in Math and Science (GEMS) title, after an extensive refinement process that includes classroom testing of trial versions, modifications to ensure the use of easy-to-obtain materials, and carefully written and edited step-by-step instructions and background information to allow presentation by teachers without special background in mathematics or science.

Staff

Glenn T. Seaborg, **Principal Investigator**
Jacqueline Barber, **Director**
Cary Sneider, **Curriculum Specialist**
Katharine Barrett, John Erickson, Jaine Kopp, Kimi Hosoume, Laura Lowell, Linda Lipner, Laura Tucker, Carolyn Willard, **Staff Development Specialists**
Jan M. Goodman, **Mathematics Consultant**
Cynthia Ashley, **Administrative Coordinator**
Gabriela Solomon, **Distribution Coordinator**
Lisa Haderlie Baker, **Art Director**
Carol Bevilacqua and Lisa Klofkorn, **Designers**
Lincoln Bergman and Kay Fairwell, **Editors**

Contributing Authors

Jacqueline Barber
Katharine Barrett
Lincoln Bergman
Jaine Kopp
Linda Lipner
Laura Lowell
Linda De Lucchi
Jean Echols
Jan M. Goodman
Alan Gould
Kimi Hosoume
Sue Jagoda
Larry Malone
Cary I. Sneider
Jennifer Meux White
Carolyn Willard

Reviewers

We would like to thank the following educators who reviewed, tested, or coordinated the reviewing of GEMS materials in manuscript form. Their critical comments and recommendations contributed significantly to these GEMS publications. Their participation does not necessarily imply endorsement of the GEMS program.

ALASKA

Olyn Garfield*
Galena City School, Galena

ARIZONA

Bill Armistead
Moon Mountain School, Phoenix

Flo-Ann Barwick
Lookout Mountain School, Phoenix

Richard E. Clark*
Washington School District, Phoenix

Bob Heath
Roadrunner School, Phoenix

Edie Helledy
Manzanita School, Phoenix

Greg Jesberger
Maryland School, Phoenix

Mark Kauppila
Acacia School, Phoenix

Karen Lee
Moon Mountain School, Phoenix

George Lewis
John Jacobs School, Phoenix

John Little
Palo Verde School, Phoenix

Tom Lutz
Palo Verde School, Phoenix

Tim Maki
Cactus Wren School, Phoenix

Don Metzler
Moon Mountain School, Phoenix

John O'Daniel
John Jacobs School, Phoenix

Donna Pickering
Orangewood School, Phoenix

Brenda Pierce
Cholla School, Phoenix

Ken Redfield
Washington School, Phoenix

Jean Reinoehl
Alta Vista School, Phoenix

Liz Sandberg
Desert Foothills School, Phoenix

Sandy Stanley
Manzanita School, Phoenix

Charri Strong
Lookout Mountain School, Phoenix

Shirley Vojtko
Cholla School, Phoenix

CALIFORNIA

Bob Alpert*
Vista School, Albany

Karen Ardito
White Hill Junior High School, Fairfax

James Boulier
Dan Mini Elementary School, Vallejo

Susan Butsch
Albany Middle School, Albany

Susan Chan
Cornell School, Albany

Robin Davis
Albany Middle School, Albany

Claudia Hall
Horner Junior High School, Fremont

Dale Kerstad*
Cave Elementary School, Vallejo

Joanna Klaseen
Albany Middle School, Albany

Margaret Lacrampe
Sleepy Hollow School, Orinda

Linda McClanahan*
Horner Junior High School, Fremont

Tina Neivelt
Cave Elementary School, Vallejo

Neil Nelson
Cave Elementary School, Vallejo

Mark Piccillo
Frick Junior High School, Oakland

Cindy Plambeck
Albany Middle School, Albany

Susan Power
Albany Middle School, Albany

Carol Rutherford
Cave Elementary School, Vallejo

Jim Salak
Cave Elementary School, Vallejo

Rich Salisbury
Albany Middle School, Albany

Secondo Sarpieri*
Vallejo City Unified School District, Vallejo

Bob Shogren*
Albany Middle School, Albany

Theodore L. Smith
Frick Junior High School, Oakland

Kay Sorg
Albany Middle School, Albany

Bonnie Square
Cave Elementary School, Vallejo

Jack Thornton*
Dan Mini Elementary School, Vallejo

Alice Tolinder*
Vallejo City Unified School District, Vallejo

Pamela Zimmerman
Cornell School, Albany

KENTUCKY

Mary Artner
Adath Jeshurun Preschool, Louisville

Alice Atchley
Wheatley Elementary School, Louisville

Sandi Babbitz
Adath Jeshurun Preschool, Louisville

Phyl Breuer
Holy Spirit School, Louisville

Toni Davidson
Thomas Jefferson Middle School, Louisville

August Drufke
Museum of History and Science, Louisville

Riva Drutz
Adath Jeshurun Preschool, Louisville

Linda Erman
Adath Jeshurun Preschool, Louisville

Jennie Ewalt
Adath Jeshurun Preschool, Louisville

Sam Foster
Museum of History and Science, Louisville

Nancy Glaser
Thomas Jefferson Middle School, Louisville

Laura Hansen
Sacred Heart Model School, Louisville

Leo Harrison
Thomas Jefferson Middle School, Louisville

Muriel Johnson
Thomas Jefferson Middle School, Louisville

Pam Laveck
Sacred Heart Model School, Louisville

Amy S. Lowen*
Museum of History and Science, Louisville

Theresa H. Mattei*
Museum of History and Science, Louisville

Brad Matthews
Jefferson County Public Schools, Louisville

Cathy Maddox
Thomas Jefferson Middle School, Louisville

Sherrie Morgan
Prelude Preschool, Louisville

Sister Mary Mueller
Sacred Heart Model School, Louisville

Tony Peake
Brown School, Louisville

Ann Peterson
Adath Jeshurun Preschool, Louisville

Mike Plamp
Museum of History and Science, Louisville

John Record
Thomas Jefferson Middle School, Louisville

Susan Reigler
St. Francis High School, Louisville

Anne Renner
Wheatley Elementary School, Louisville

Ken Rosenbaum
Jefferson County Public Schools, Louisville

Edna Schoenbaechler
Museum of History and Science, Louisville

Melissa Shore
Museum of History and Science, Louisville

Joan Stewart
DuPont Manual Magnet School, Louisville

Jenna Stinson
Thomas Jefferson Middle School, Louisville

Dr. William M. Sudduth*
Museum of History and Science, Louisville

Larry Todd
Brown School, Louisville

Harriet Waldman
Adath Jeshurun Preschool, Louisville

Fife Scobie Wicks
Museum of History and Science, Louisville

August Zoeller
Museum of History and Science, Louisville

Doris Zoeller
Museum of History and Science, Louisville

MICHIGAN

Dave Bierenga
South Christian School, Kalamazoo

Edgar Bosch
South Christian School, Kalamazoo

Craig Brueck
Schoolcraft Middle School, Schoolcraft

Joann Dehring
Woodland Elementary School, Portage

Tina Echols
Lincoln Elementary School, Kalamazoo

Barbara Hannaford
Gagie School, Kalamazoo

Dr. Alonzo Hannaford*
Science and Mathematics Education Center
Western Michigan University, Kalamazoo

Rita Hayden*
Science and Mathematics Education Center
Western Michigan University, Kalamazoo

Mary Beth Hunter
Woodland Elementary School, Portage

Ruth James
Portage Central High School, Portage

Dr. Phillip T. Larsen*
Science and Mathematics Education Center
Western Michigan University, Kalamazoo

Gloria Lett*
Kalamazoo Public Schools, Kalamazoo

Roslyn Ludwig
Woodland Elementary School, Portage

David McDill
Harper Creek High School, Battle Creek

Everett McKee
Woodland Elementary School, Portage

Susie Merrill
Gagie School, Kalamazoo

Rick Omilian*
Science and Mathematics Education Center
Western Michigan University, Kalamazoo

Kathy Patton
Northeastern Elementary School, Kalamazoo

Rebecca Penney
Harper Creek High School, Battle Creek

Shirley Pickens
Schoolcraft Elementary School, Schoolcraft

Deb Ply
South Junior High School, Kalamazoo

Sue Schell
Gagie School, Kalamazoo

Sharon Schillaci
Schoolcraft Elementary School, Schoolcraft

Julie Schmidt
Gagie School, Kalamazoo

Joel Schuitema
Woodland Elementary School, Portage

Bev Wrubel
Woodland Elementary School, Portage

NEW YORK

Frances Bargamian
Trinity Elementary School, New Rochelle

Bob Broderick
Trinity Elementary School, New Rochelle

Richard Golden*
Webster Magnet Elementary School, New Rochelle

Tom Mullen
Jefferson Elementary School, New Rochelle

Edna Neita
George M. Davis Elementary School, New Rochelle

Sigrin Newell
Discovery Center, Albany

Eileen Paolicelli
Ward Elementary School, New Rochelle

Dr. John V. Pozzi*
City School District of New Rochelle, New Rochelle

John Russo
Ward Elementary School, New Rochelle

Bruce Seiden
Webster Magnet Elementary School, New Rochelle

David Selleck
Albert Leonard Junior High School, New Rochelle

Tina Sudak
Ward Elementary School, New Rochelle

Julia Taibi
George M. Davis Elementary School, New Rochelle

Rubye Vester
Columbus Elementary School, New Rochelle

Bruce Zeller
Isaac E. Young Junior High School, New Rochelle

NORTH CAROLINA

Jorge Escobar
North Carolina Museum of Life and Science, Durham

Ed Gray
Discovery Place, Charlotte

Sue Griswold
Discovery Place, Charlotte

Mike Jordan
Discovery Place, Charlotte

James D. Keighton*
North Carolina Museum of Life and Science, Durham

Paul Nicholson
North Carolina Museum of Life and Science, Durham

John Paschal
Discovery Place, Charlotte

Cathy Preiss
Discovery Place, Charlotte

Carol Sawyer
Discovery Place, Charlotte

Patricia J. Wainland*
Discovery Place, Charlotte

OHIO

A.M. Sarquis
Miami University, Middletown

OREGON

Christine Bellavita
Judy Cox
David Heil*
Shab Levy
Joanne McKinley
Catherine Mindolovich
Margaret Noone*
Jim Todd
Ann Towsley
Oregon Museum of Science and Industry

Oregon Museum of Science and Industry (OMSI) staff conducted trial tests at the following sites:
Berean Child Care Center, Portland
Grace Collins Memorial Center, Portland
Mary Rieke Talented and Gifted Center,
   Portland Public School District, Portland
Portland Community Center, Portland
Portland Community College, Portland
St. Vincent De Paul, Child Development
   Center, Portland
Salem Community School, Salem
Volunteers of America, Child Care Center,
   Portland

WASHINGTON

David Foss
Stuart Kendall
Dennis Schatz*
William C. Schmitt
David Taylor
Pacific Science Center, Seattle

FINLAND

Sture Björk
Åbo Akademi, Vasa

Arja Raade
Katajanokka Elementary School,
Helsinki

Pirjo Tolvanen
Katajanokan Ala-Aste,
Helsinki

Gloria Weng*
Katajanokka Elementary School,
Helsinki

*Trial test coordinators

Wonder and curiosity abound at a Bubble Festival learning station. See the note about the GEMS guide Bubble Festival *on page 1 of this guide.*

Contents

"Blow a soap bubble and observe it. You may study it all your life and draw one lesson after another in physics from it."

— *Lord Kelvin*

Acknowledgments

These bubble activities were first developed by Jacqueline Barber and Nancee Boice. Alan Friedman, Cary Sneider, and Susan Jagoda assisted in further development of "Comparing Bubble Solutions" and "Predict-A-Pop." Earlier published versions of these latter two activities appeared in "Doing Science," sponsored by the American Association for the Advancement of Science, 1983.

Introduction

Watching a soapy puddle turn into a beautiful, fragile, multi-colored sphere is always a fascinating experience. This fascination accounts for the universal and eager interest in bubbles.

Bubbles are not only captivating, colorful, and fun to make, they are also excellent demonstrations of scientific phenomena. Some of the topics introduced in this booklet are:

- light and color
- aerodynamics
- chemical composition
- surface tension
- technology

Your students will also have an opportunity to develop a variety of skills useful for continued study in science and for daily life. These skills include:

- making observations
- measuring and recording data
- experimenting
- calculating averages
- graphing results
- making inferences
- drawing conclusions

A few things to know before you get started:

Each activity introduces concepts and skills that build upon what students learn in previous activities. The early activities allow students to blow bubbles to their hearts' content. Giving free rein to this very natural impulse helps pave the way for later activities, which require that students blow bubbles and then *not* pop them.

Because bubbles are so interesting and exciting, some teachers prefer to present activities over a longer time period, rather than concentrating them in a one or two week period.

Bubble-ology is a motivating and powerful introduction to the process and substance of science... and besides, it's good, clean, fun. So roll up your sleeves and get started!

The GEMS series includes another teacher's guide focusing on bubbles, Bubble Festival, which explains how to set up classroom table-top learning stations to do bubble activities. Written for Grades K–6, Bubble Festival learning station activities can be adapted for all ages, and many successful large-group Bubble Festivals have been held for older students and adults. There are a number of stations in the Bubble Festival that are not part of Bubble-ology. Even where some overlap exists, each guide provides the student with a different and quite unique learning experience. For example, "Bubble Colors" in Bubble Festival, in which children discover different colors and patterns, can lead nicely to "Predict-a-Pop" in Bubble-ology, in which students record the sequence of colors they observe and make generalizations about their observations. Similarly, in the "Bubble Measurement" festival station children discover that bubbles can be measured. In Bubble-ology, students use measurement as part of a controlled experimental process to compare different brands of soap solution. If you choose to use both guides, keep in mind that, for older students, the Bubble Festival could provide an excellent introduction to the more formal, structured, guided discovery activities of Bubble-ology.

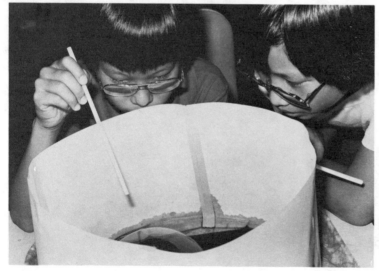

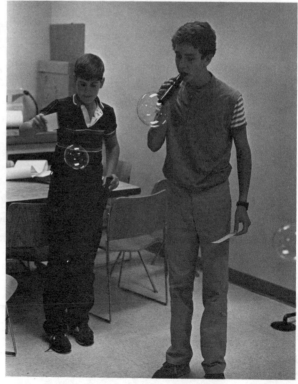

Time Frame

Activity 1: Bubble Technology
    Teacher Preparation: 20 minutes
    Classroom Activity: one or two 45-minute
        sessions

Activity 2: Comparing Bubble Solutions
    Teacher Preparation: 20 minutes
    Classroom Activity: one 40- to 60-minute
        session

Activity 3: The Chemistry of Bigger Bubbles
    Teacher Preparation: 15 minutes
    Classroom Activity: two 45-minute sessions

Activity 4: Bernoulli's Bubbles
    Teacher Preparation: 15 minutes
    Classroom Activity: one 45-minute session

Activity 5: Predict-A-Pop
    Teacher Preparation: 15 minutes
    Classroom Activity: one or two 45-minute
        sessions

Activity 6: Longer Lasting Bubbles
    Teacher Preparation: 20 minutes
    Classroom Activity: two 45- to 60-minute
        sessions

Activity 1: Bubble Technology

Introduction

Can you blow a bubble with a pair of scissors, a piece of paper, or a rubber band? In this activity, your students experiment to discover what objects can be used to blow bubbles, which make little bubbles, and which make big bubbles. Then they use the information they've gathered to design and draw bubble-makers for specialized uses. This activity introduces students to the process of technology and allows them to get some of the bubble blowing out of their systems—something you'll appreciate during later activities!

What You Need

For preparation and cleanup:
- [ ] newspapers to cover tables
- [ ] 8 oz. (240 ml) dishwashing liquid
- [ ] water
- [ ] measuring cup or graduated cylinder
- [ ] eyedropper
- [ ] 1 one-gallon container for mixing bubble solution
- [ ] glycerin (optional)

Complete classroom kits for GEMS teacher's guides are available from Sargent-Welch. For further information call 1-800 727-4368 or visit www.sargentwelch.com

For the class:
- [ ] at least ten different materials to use for bubble-makers, such as: strainer, small tin cans, protractors, paper, mason jar lids, string, drinking straws, tea ball, rubber stoppers with holes, flower pots, funnels, eyedroppers, turkey basters, rubber tubing, paper cups, styrofoam cups, various gauges of screen, different sized washers, rubber bands, toilet-tissue and paper-towel rolls, aluminum foil, wire of different gauges, springs, scissors, tubes of any kind, oatmeal box, and anything else you think appropriate.

Consider asking students to bring in possible bubble-maker materials from home.

For each group of 3–4 students:
- [ ] 1 wide-mouthed, flat-bottomed pan (such as a metal pie pan, dish pan, or other container suitable for holding bubble solution)

If student sharing of the same bubble blowing instrument is considered a problem, some teachers have suggested providing a bleach rinse between uses.

Getting Ready

1. Prepare one gallon of bubble solution:
   1 cup (240 ml) dishwashing liquid
   50–60 drops glycerin (optional)
   1 gallon water (3.8 liters)

2. Cover tables with newspapers.

3. Fill containers with about 2 cups of bubble solution and set them on tables.

4. Place all bubble-maker materials you have been able to collect on a central table.

5. Clear two other tables or surfaces on which students will place bubble-maker materials after they've been tested—one table for those that work, and one for those that do not.

Presenting The Challenge

1. Ask: "Who has blown soap bubbles before?" "What did you use to blow the bubbles?" Explain that the challenge during this session is to discover what materials or objects can be used to blow bubbles and what kinds of bubbles these materials make.

2. Point out the materials to experiment with to the students. Tell them that they should each test 8 to 10 different materials.

3. After students test a material, they should place it on one of two tables—one for materials that succeed in making bubbles, and one for materials that do not. Point out that any student may try materials from the "don't work" table over again. If this new attempt succeeds, that material should be moved to the table for materials that work.

4. Write "small" on one side of the board and "large" on the opposite side. Ask the students to classify each material they test as to whether it makes large or small bubbles. Have them write the names of the objects on the board, in order according to the bubble size they produce.

Experimenting

As students experiment, circulate among them suggesting questions to investigate, such as:

- What would happen if you changed the shape of the wire?

- Does the length of the tube make a difference?

- Which kind of cup works better—paper or styrofoam?

- Can you think of a way to use this paper towel to blow a bubble?

Drawing Conclusions

1. Gather the students around the chalkboard. Ask:

- "What's the same about all working bubble-makers?"

- "What objects didn't work?"

- "What were these objects lacking?"

- "What strategies did you use to make something work as a bubble-maker?"

2. As a group, go over the list of bubble-makers according to bubble size. What characteristics did makers of smaller bubbles share? Makers of larger bubbles?

3. Explain that *technology* involves the use of science to create something practical. Point out to your students that they have just conducted a series of experiments to gather information about bubble-blowing tools. Now you'd like them to use that information to create a practical application of science. Creating practical applications of science is what engineers do.

Give your students a homework challenge: to create imaginative drawings of bubble-makers for specialized uses. For example, a bubble-maker that: creates foam; doesn't need to be dipped into a soap solution; makes large, detachable bubbles; or other creative ideas your students might suggest.

Going Further

Have your students combine materials to make more complicated bubble-making contraptions.

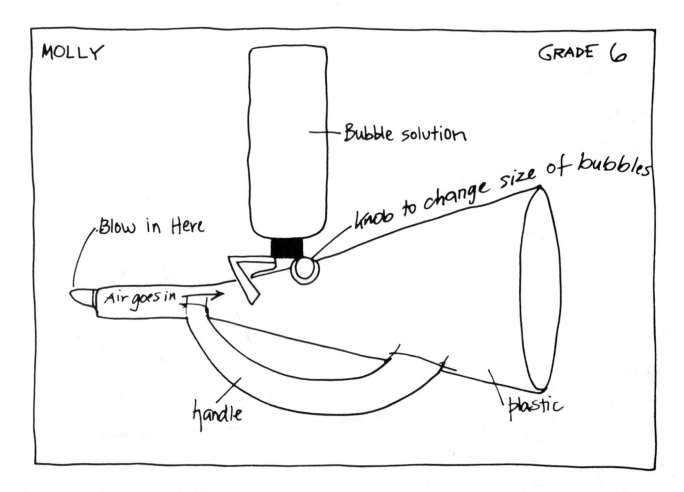

Activity 2: Comparing Bubble Solutions

Introduction

Challenge your students to determine which brand of dishwashing liquid will make the biggest bubble. This activity presents your students with a way to quantify how well a soap solution forms bubbles. It also introduces the scientific concept of a *fair test.*

What You Need

For preparation and cleanup:

- [ ] 8 oz. (240 ml) of three different brands of dishwashing liquid (include one cheap and one expensive)
- [ ] water
- [ ] 1 measuring cup or graduated cylinder
- [ ] 1 eyedropper
- [ ] 3 one-gallon containers for mixing bubble solution
- [ ] 1 roll of masking tape
- [ ] paper towels
- [ ] 2 cups vinegar
- [ ] glycerin (optional)
- [ ] squeegie (optional)

For each pair of students:

- [ ] 1 meter or yard stick
- [ ] 2 plastic drinking straws
- [ ] 1 one-pint container (such as a cottage cheese container) for holding bubble solution
- [ ] 1 "Bubble Solutions" data sheet (master included, page 16)
- [ ] 1 pencil
- [ ] 1 table, counter, desk, or board about 30" (75 cm) in diameter
- [ ] calculator (optional)

Getting Ready

1. Make one copy of the "Bubble Solutions" data sheet for each pair of students.

2. Label one of the large and a third of the small containers with the name of each dishwashing liquid.

3. Prepare three different bubble solutions (one bubble solution from each brand of liquid) in the three large containers:

> 1 cup (240 ml) dishwashing liquid
> 50–60 drops glycerin (optional)
> 1 gallon water (3.8 liters)

4. Pour about one cup of the appropriate solution into each of the small, labeled containers.

5. For each pair of students, set up one test station on a flat surface, about 30" (75 cm) in diameter. Place a small container of one type of solution and a meter stick at each station.

Introducing the Activity

1. Explain that the object of this session is to compare brands of dishwashing liquid to find out which solution makes the biggest bubbles. Ask the students for their ideas on how they might design an experiment to make this comparison.

2. After several ideas for experiments have been shared, explain, if it has not already become clear, that in order to compare dishwashing liquids as to bubble size, a standard method to measure the bubbles is needed. Demonstrate the following procedure for measuring bubbles:

   a. Pour some soap solution on the surface of the table, and use your hand to wet an area about 18" (45 cm) in diameter.

   b. Dip a straw into the solution in the container.

c. With the straw just touching the soapy surface of the table, gently blow through the straw to form a bubble dome, and continue blowing until it pops. Take more than one breath if necessary.

d. With a meter stick, show the students how to measure the inside diameter of the ring of soap suds left by the bubble dome.

Tell your students that using this method provides one standard way to measure the size of a bubble.

Experimenting

1. Have the students work in pairs to measure bubbles from all three solutions. Suggest that they alternate blowing bubbles to reduce the chance of hyperventilation—students have been known to blow big bubbles! Instruct each pair of students to measure four bubbles per brand and average their results. Then have them move on to a solution with another brand of soap and again measure four bubbles.

2. As your students start to experiment, they will come to you with questions such as: "He popped my bubble. Does that bubble have to count?" "This bubble is oblong. How do I measure its diameter?" You may also notice students who won't record the diameter of smaller bubbles. Respond by saying that they are the scientists. It's up to them to decide what is fair. Whatever they decide to do, they'll need to make sure all three brands are treated in the same way. Otherwise one brand of soap will be given an unfair advantage.

If your students have difficulty calculating averages, use the following method to analyze the results:

—Have your students write all of their bubble dome measurements on the board.

—For each brand, count the number of bubble domes that were larger than a given diameter (e.g., 30 cm).

—Use this number to compare brands and determine which brand made bigger bubbles.

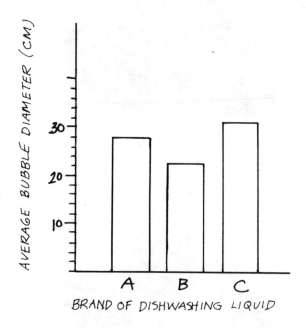

AVERAGE BUBBLE DIAMETER (CM)

BRAND OF DISHWASHING LIQUID

Analyzing Results

1. Reconvene the class around the chalkboard. Write the names of the dishwashing liquids across the top of the board, and record the students' averages under each column. Ask your students to calculate the grand average for each brand of soap solution.

2. As a group, make a bar chart on the chalkboard, comparing the average bubble diameter for each soap brand. Rank the brands of soap from biggest to smallest bubble-maker.

3. Ask the students if this has been a "fair test." In testing the three brands of soap solution: Was the same amount of bubble solution used each time? Were all bubbles measured in the same way? Was there anything else that might have affected the results?

Explain that the factors that change from one time a bubble is blown to the next are called *variables*. Point out that in a fair test, all variables are kept the same or "controlled" except for the *test variable*. Scientists call this a *controlled experiment*. Ask what they think the test variable was in this experiment. [The brand of soap.]

4. See if your students have any ideas why certain dishwashing liquids make bigger bubbles than others [e.g. the proportion of soap to water, the amount and type of additives used, etc.]. Tell them that in the next activity, "The Chemistry of Bigger Bubbles," they will investigate *why* some substances are better bubble-makers than others.

Cleanup

To clean soapy tables: first use a squeegie or paper towels to remove excess bubble solution. **Do not add water.** Then sprinkle vinegar on the area to cut the soap film. Wipe dry with paper towels. Repeat once more if surface still retains soap film.

Going Further

1. Challenge your students to devise ways to test additional bubble qualities (e.g. life span, durability).

2. Ask your students how they might design an experiment that tests the dishwashing properties of various brands of dishwashing liquid.

3. Show your students how to express their results as the average *volume* of a bubble dome, rather than an average diameter. Assume that a bubble dome is half of a sphere. Determine the length of the bubble dome radius (r), and calculate the volume (v) of a sphere:

$$v = 4/3 \; \pi \; r^3$$

Then divide by two to determine the volume of half of a sphere.

Make further use of the formula for the volume of a sphere by having the class calculate their own lung volumes. Instruct each student to blow a bubble dome with three lungfuls of air. Have them determine the volume of that bubble dome and divide by three to find their average lung volume.

BUBBLE SOLUTIONS

NAME(S):_____

Soap Brand_____

Bubble 1_____ 2_____

3_____ 4_____

Average Bubble Diameter_____

Soap Brand_____

Bubble 1_____ 2_____

3_____ 4_____

Average Bubble Diameter_____

Soap Brand_____

Bubble 1_____ 2_____

3_____ 4_____

Average Bubble Diameter_____

WHICH SOAP BRAND HAD THE BIGGEST AVERAGE BUBBLE DIAMETER?

Activity 3: The Chemistry of Bigger Bubbles

Introduction

Why do some dishwashing liquids make bigger bubbles than others? Why does cream form bubbles when it is whipped, while milk does not? An enormous variety of natural substances form bubbles. Sea foam is formed by the agitation of phosphates (like those in soaps) released by decomposing kelp. Egg whites form hundreds of tiny bubbles when beaten. In each case, the formation of bubbles depends on the *chemical composition* of the substance.

This activity introduces your students to some of the properties of bubble-making substances. The students observe how soap affects the *surface tension* of water and investigate the role of evaporation in bubble formation, as they test the effect of different amounts of glycerin on the size of bubbles.

What You Need

For preparation and cleanup:

- [ ] 8 oz. (240 ml) dishwashing liquid
- [ ] water
- [ ] measuring cup or graduated cylinder
- [ ] 1 one-gallon container for mixing bubble solution
- [ ] 1 roll of masking tape
- [ ] paper towels
- [ ] 2 cups vinegar
- [ ] 1 squeegee (optional)

For the class:

- [ ] several ounces of glycerin
- [ ] several eyedroppers
- [ ] several measuring cups
- [ ] several calculators (optional)
- [ ] chalkboard
- [ ] chalk

You may choose to measure the varying drops of glycerin that will be needed for the experiment into different clear containers before class. This can greatly lessen the time taken in class to add the glycerin drop by drop. The clear containers allow students to see that different quantities are being added.

For each pair of students:

- ☐ 1 meter or yard stick
- ☐ 2 plastic drinking straws
- ☐ 1 one-pint container (such as a cottage cheese container) for holding bubble solution
- ☐ 1 "Experimenting with Glycerin" data sheet (master included, page 26)
- ☐ 1 graphing sheet (master included, page 27)
- ☐ 1 pencil
- ☐ 1 table, counter, desk, or board about 30" (75 cm) in diameter

For the demonstration:

- ☐ 1 tall, clear, drinking glass
- ☐ water
- ☐ water pitcher
- ☐ 1 eyedropper
- ☐ dishwashing soap (just 1 drop)

Getting Ready

1. Make one copy of the "Experimenting with Glycerin" data sheet and of the graphing sheet for each pair of students.

2. Prepare a gallon of bubble solution without glycerin:

    1 cup (240 ml) dishwashing liquid
    1 gallon water (3.8 liters)

3. Clear one flat surface, about 30" (75 cm) in diameter, for each pair of students.

4. Place the demonstration materials on a table or desk.

5. Label eight pint containers "A" through "H." Fill all eight containers with one cup of bubble solution made with no glycerin. Leave container "A" without glycerin. Add 10 drops of glycerin to container "B," 20 drops to container "C," and so on through container "H," which will have 70 drops of glycerin. Hold the eyedropper **vertically** in order to help reduce variation in size of drops.

6. Place these containers of bubble solution and all other materials on a centrally located table.

Observing Surface Tension

1. Ask your students what substances they can think of that form bubbles. Point out that some substances, such as water, do form bubbles, but these bubbles disappear almost as quickly as they are formed.

2. Perform the following demonstration to explain why pure water bubbles don't last:

- Gather the group around the demonstration table. Have them squat down so their eye level is closer to the level of the table.

- Fill a glass to the top with water. Keep adding water, drop by drop, until you think the glass will overflow. Then add a few more drops. If you are careful, you'll be able to add water until the surface of the water is actually higher than the glass.

- Ask students if they can see that the water behaves as if it were covered with a skin. Explain that this effect is called *surface tension*. Water molecules at the surface of water are more attracted to each other than to the air; it is as if they stick together. This "stickiness" causes surface tension. Surface tension keeps water from spilling and discourages the formation of bubbles. When bubbles do form, they are short-lived.

If you have an extra twenty minutes and access to enough eyedroppers for each student, consider replacing this teacher demonstration with a hands-on student activity. Give each student an eyedropper and a penny. Ask them to predict how many drops of water will fit on the penny without spilling. Distribute dishes of water and have them find out.

After the students have a chance to observe the surface tension of the water on a penny, ask them to put drops of water on a penny again. Then have them "break" the surface tension of the water on the penny by adding a drop of soap solution.

- To demonstrate the effect of soap on surface tension, carefully add one drop of soap to the very full glass of water. This should "break" the surface tension of the water, causing it to overflow. Explain that soap decreases the surface tension of water to about one-third of what it usually is: just right for making bubbles.

Discussing the Problem of Evaporation

1. Point out that another problem with using water to blow bubbles is that water evaporates very rapidly. When the water evaporates, the bubble wall is broken. This problem is not limited to the use of pure water since most soap bubble solutions contain water.

2. Explain that scientists have devised a way to deal with the problem of evaporation: adding a substance to the bubble solution to keep water from evaporating. Substances that have water-holding properties are referred to as *hygroscopic*. Glycerin is a hygroscopic liquid that is typically added to bubble solutions. Glycerin forms a weak chemical bond with water that delays evaporation.

Planning the Experiment

1. Tell your students that their challenge is to determine what effect the *amount* of glycerin in a bubble solution has on the *size* of the bubbles formed.

2. Ask the students for their ideas on ways of designing the experiment so it is a fair comparison. Use the following questions to guide the group in determining the test procedure they'd like to use.

- What is the test variable? [Amount of glycerin.]
- What variables must be kept the same or "controlled"?

3. Present a plan for varying the amount of glycerin while keeping the amount of water and dishwashing liquid the same. Draw eight cups on the chalkboard. Tell the class that each test formula will start with one cup of bubble solution made without glycerin. Formula A will have 0 drops of glycerin added to it. Formula B will have 10 drops of glycerin added to it, and so on up to Formula H which will have 70 drops of glycerin added. On the chalkboard, record the letter of the formula and the number of glycerin drops in each "cup."

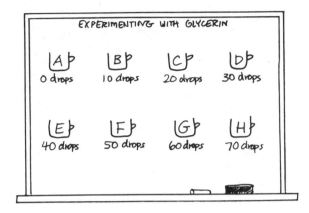

4. Ask the students if they have any expectations about the experiment. How much glycerin do they think will make the biggest bubbles?

Experimenting

1. Assign pairs of students to test the formulas. Have them apply the same method for measuring bubble size used in Activity 2: Comparing Bubble Solutions, as explained on pages 12 and 13.

2. As students finish testing one formula, have them swap work stations and test other formulas. For best results, each formula should be tested by at least four different groups.

Graphing the Results

1. Ask your students to gather around the chalkboard. Record the students' averages under the formula names written across the top of the board. Calculate the grand average for each formula.

2. Ask your students to graph the results of all experiments on a graphing sheet (master included). Does the graph show an optimum amount of glycerin for making the biggest bubbles?

Note: Sometimes student data has so much variation that it is difficult to identify an optimum amount of glycerin. If so, have your group identify a broader range for the desirable amount of glycerin. Ask them how they would improve the experiment in order to pinpoint more exactly the optimum amount of glycerin.

3. You may want to ask your students if they were surprised by the results. Many students assume at first that the more glycerin used, the bigger the bubbles will be. As this experiment demonstrates, that is not the case.

Going Further

1. Sugar is another hygroscopic substance. Challenge your students to repeat their experiments using sugar instead of glycerin. Compare the results of the two experiments. Which is better for making big bubbles, sugar or glycerin?

2. Give your students the open-ended challenge of developing their own ideal bubble solutions. Remind them to vary only one ingredient at a time as they experiment, and to keep a careful record of what they do.

EXPERIMENTING WITH GLYCERIN

NAME(S): _____

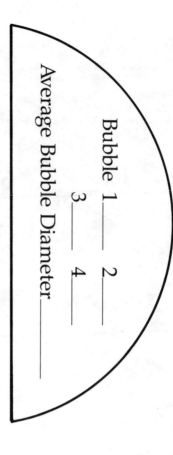

Bubble 1 ___ 2 ___
3 ___ 4 ___

Average Bubble Diameter _____

Number of Drops of Glycerin _____

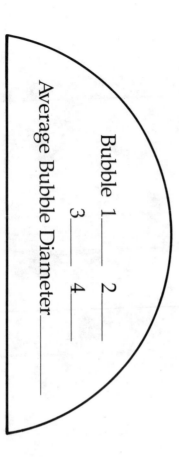

Bubble 1 ___ 2 ___
3 ___ 4 ___

Average Bubble Diameter _____

Number of Drops of Glycerin _____

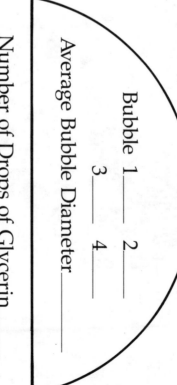

Bubble 1 ___ 2 ___
3 ___ 4 ___

Average Bubble Diameter _____

Number of Drops of Glycerin _____

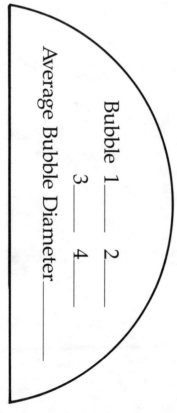

Bubble 1 ___ 2 ___
3 ___ 4 ___

Average Bubble Diameter _____

Number of Drops of Glycerin _____

GRAPHING SHEET

*NAME(S):*_____

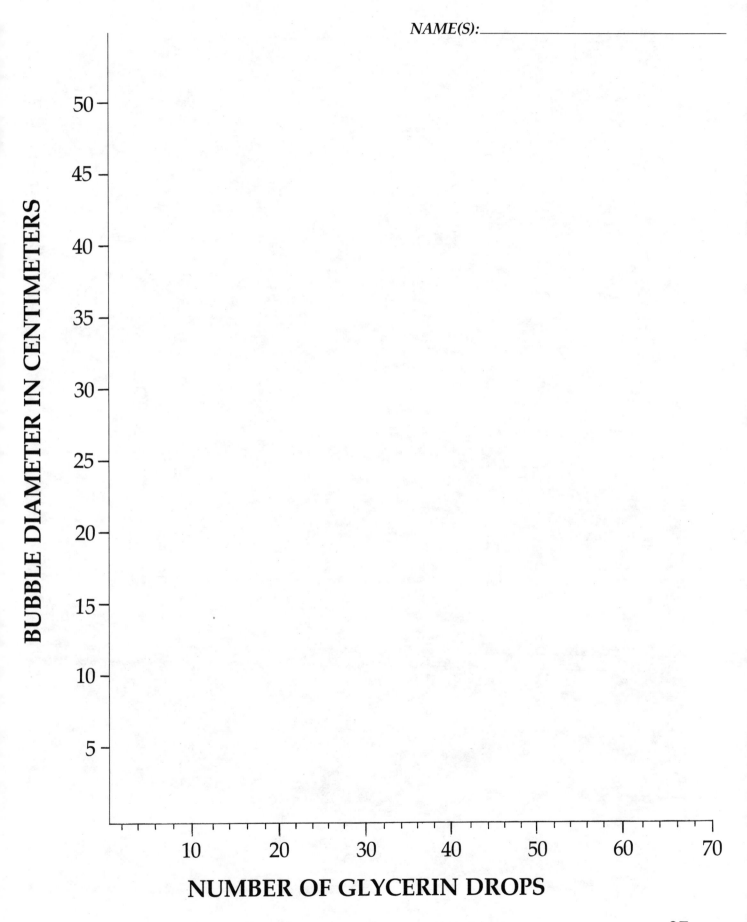

BUBBLE DIAMETER IN CENTIMETERS

NUMBER OF GLYCERIN DROPS

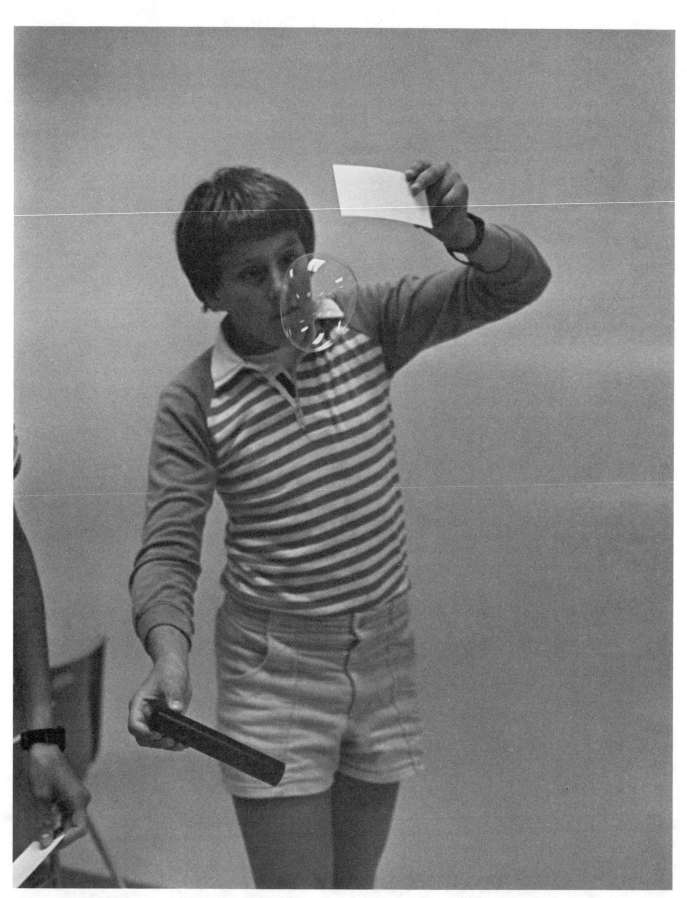

Activity 4: Bernoulli's Bubbles

Introduction

For centuries, scientists have been interested in finding ways to make things fly. From the ancient Chinese, who made large kite-gliders, to Leonardo da Vinci, who designed a machine with flapping wings, humanity pursued the dream of flight.

In the 18th century, a scientist named Daniel Bernoulli discovered a scientific principle that now carries his name. Although he was working with liquids, the principle can be applied in many ways, and it became the basis for airplane flight many years after its discovery.

In this activity, you'll introduce aerodynamics to your students by challenging them to devise the best ways to keep a bubble aloft. In this fun context, you'll teach Bernoulli's principle and help explain how airplanes fly.

What You Need

For preparation and cleanup:
- ☐ newspapers to put under containers of bubble solution
- ☐ 8 oz. (240 ml) dishwashing liquid
- ☐ water
- ☐ 1 measuring cup or graduated cylinder
- ☐ 1 eyedropper
- ☐ 1 one-gallon container for mixing bubble solution
- ☐ glycerin (optional)

For each group of 4–6 students:
- ☐ 1 pint-sized container (such as a cottage cheese container) for holding bubble solution

For each student:
- ☐ 1 tube (about 7"–11" in length, 1"–2" in diameter—such as: plastic golf club covers cut in 7" lengths, cardboard paper towel rolls, polyvinylchloride pipes, or two small cans taped end-to-end with lids removed)
- ☐ 1 3"x5" index card

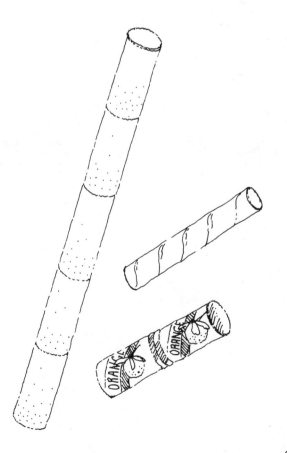

Getting Ready

1. Prepare one gallon of bubble solution:
   1 cup (240 ml) dishwashing liquid
   50–60 drops glycerin (optional)
   1 gallon water (3.8 liters)

2. Place small containers of the bubble solution on newspaper, in various locations around the room.

3. Spend five minutes practicing the technique of keeping a bubble in the air by waving your hand back and forth over it. You will want to master this technique before you demonstrate it for the students.

Hints:

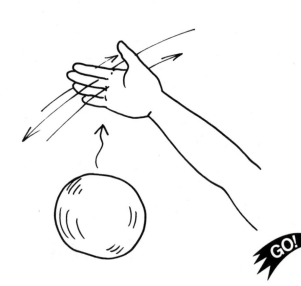

- If you have trouble detaching the bubble from the bubble-blower, try flicking your wrist toward you so the end of your bubble-blower points upward.

- If you are having trouble keeping your bubble in the air, try this: dip your blower in solution; blow and release one bubble; blow and release a second bubble. Try keeping the second bubble up—"second bubbles" are lighter because they are made with less bubble solution.

GO! ➤ Exploring

1. Challenge your students to devise ways to keep a bubble from hitting the ground, without touching it with their hands or with any other objects. As they find methods that work, encourage them to experiment with other approaches. If students have difficulty detaching the bubbles from their bubble-blowers, show them how to flick their wrists toward themselves so the end of the bubble-blower points upward.

2. Have the students set their bubble-blowers aside and sit where they can see the chalkboard. As a group, make two lists: methods that worked, and those that didn't work.

Explaining

1. Present the brief historical perspective given in the first paragraph of this activity. Explain that the *Bernoulli principle* states that the faster air flows, the less pressure it exerts.

2. Draw a diagram of an airplane and an airplane wing on the chalkboard. Point out that as air hits the wings of a plane, some of it has to go over them and some of it has to go underneath. Scientists have discovered that regardless of whether the air goes over or under, it arrives at the other side of the wing at the same instant. The same holds true for the air that goes over and under the body of the plane. Ask your students the following questions:

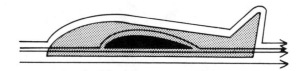

- Which air has to travel faster to make it to the other side; the air that goes over the wing or under the wing? [The air that goes over the wing: it has farther to travel in the same time period.]

- What does the Bernoulli principle say about faster moving air? [It exerts less pressure.]

- If there is greater pressure pushing up from under the airplane, what will happen? [The airplane will go up!]

3. Explain that the force pushing upward is called *dynamic lift*. You might add that other engineering principles are brought into play to provide additional lift.

Applying the Principle

1. Show how to use Bernoulli's principle to decrease the pressure over a bubble. Demonstrate how you can keep a bubble in the air by waving your hand back and forth over it. Waving your hand over the bubble causes the air to move faster; faster moving air exerts less pressure. The faster you wave, the higher the bubble goes.

2. Challenge your students to use Bernoulli's principle to keep their bubbles aloft. Make sure you allow enough time for all students to be successful. If students have difficulty, give them the hint about using the second bubble described on page 30 in the "Getting Ready" section.

3. After the students have time to try the hand-waving method, distribute index cards and invite them to wave the cards to further demonstrate Bernoulli's principle.

Going Further

1. Set up a series of short bubble obstacle courses, including challenging features such as steps, curves, corners, and a hoop. After your students have had time to try maneuvering their bubbles through the course, lead a discussion: What was the most difficult situation? What refinements did they develop to increase the amount of distance the bubbles were able to travel through the course?

2. Have your students research Daniel Bernoulli and the history of flight.

3. Make or obtain posters of airplanes and airplane wings, and post them around the classroom. Ask students to explain how the Bernoulli principle is incorporated into the design of each plane.

4. Challenge students to use bubbles to detect air flow patterns in a room or outdoors. In which direction is the wind blowing? Does that air vent blow air into the room? What happens to currents near the building? Can they observe any whirlpool movement? Provide them with a map or help them create one. Discuss techniques they can use to indicate direction of air flow on their maps.

Propelling a bubble through an obstacle course requires not only that the bubble stay aloft, but also that it move in a forward direction. If students don't figure this out for themselves, suggest that they work in pairs, having one person wave over the bubble and the other person wave behind the bubble.

Activity 5: Predict-A-Pop

Introduction

Blow a soap bubble. Can you tell when it will pop? You and your students may have already discovered that color is one important clue. It's interesting that color should be a key to predicting bubble survival, since we usually think of color as a mere surface decoration. But actually the colors of a soap bubble are produced by a complex interaction between light and matter called *interference*.

This activity is a playful introduction to interference, an important phenomenon in the history of physics and in modern industry. Your students will enjoy discovering how to count down the last few seconds of a bubble's existence..
3..2..1... POP!!

The December 1995 (Volume 33, Number 9) issue of The Physics Teacher, from the American Association of Physics Teachers, carries an excellent article entitled "Videotaping the Lifespan of a Soap Bubble" by Göran Rämme of Uppsala Sweden describing how a video camera can be used to document a great deal of interesting soap-film phenomena. The magazine's cover contains this quotation from Lord Kelvin: "Blow a soap bubble and observe it. You may study it all your life and draw one lesson after another in physics from it."

What You Need

For preparation and cleanup:
☐ 8 oz. (240 ml) dishwashing liquid
☐ water
☐ 1 measuring cup or graduated cylinder
☐ 1 eyedropper
☐ 1 one-gallon container for mixing bubble solution
☐ several rolls of masking tape
☐ glycerin (optional)

Black or dark brown plastic garbage bags can also be used to provide the dark surface contrast needed in this activity. Cut the trash bag along one side and the bottom crease so it opens out in one flat piece. Make sure to first sprinkle a few drops of water on the surface to be covered, then lay the trash bag down, black side up, and smooth it down. The slight dampness helps the plastic adhere to the surface.

For each pair of students:
☐ 1 pint-sized container for holding bubble solution
☐ 2 plastic drinking straws
☐ 6 8½"x11" sheets of white paper
☐ 1 flat, dark surface about 18" (45 cm) in diameter
        or
☐ 1 cafeteria tray and black construction paper to cover the tray

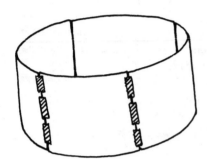

Getting Ready

1. Prepare one gallon of bubble solution:
   1 cup (240 ml) dishwashing liquid
   50–60 drops glycerin (optional)
   1 gallon water (3.8 liters)

2. Pour the bubble solution into the small containers. Place the containers in a central location along with straws, white paper, and masking tape.

3. Clear off a flat, dark surface (about 18" [45 cm] in diameter) for each pair of students.

4. Prepare one "white collar" by taping four sheets of white paper together so they form a cylinder 8½" high. The white collar reduces air currents and reflects light onto the bubble so its colors can be seen clearly.

GO! ▶ ## Observing Colors

1. Gather the students around you. Blow a bubble dome as follows:

   a. Pour about ⅓ cup of soap solution on the surface of the table or tray, and use your hand to wet an area about 18" (45 cm) in diameter.

   b. Place the white collar around the soapy area.

   c. Dip a straw into the soap solution.

   d. With the straw just touching the surface of the table, gently blow through the straw to form a bubble dome.

   e. Remove the straw.

If your students have difficulty seeing the colors on top of the bubble, suggest that they position a piece of white paper so that it will reflect more light onto the top of the bubble.

2. Explain that the challenge for the day is to use color to recognize the moment just before a bubble pops. Instruct each pair of students to start by making a collar, blowing a bubble dome, and observing the changing colors *on top of the bubble*. Tell them to record the sequence of colors they see for four or five bubbles.

Reporting Results

Have the students leave their materials and form a circle in view of the chalkboard. As several of the teams report, record their findings on the board. The students will probably discover a repeating sequence something like this: green to blue to magenta to yellow to green... (sequence repeats more than once)... and finally white to white with black spots to black—POP!! (The spots are actually transparent but because the background is black, they appear black.) Explain that the colors on the surface of a bubble change as the bubble becomes thinner and thinner.

Not all students will see this pattern because air currents may interfere with the gradual thinning of the top of the bubble, interrupting the usual color sequence. Write the typical color sequence on the board and draw the decreasing bubble wall underneath it. (See diagram.) Explain that in cases where there are absolutely no air disturbances, such as a bell jar, this is the pattern scientists have reported seeing. Ask the students if they notice any aspects of the typical pattern in their data.

GREEN-BLUE-MAGENTA-YELLOW-GREEN ...(SEQUENCE REPEATS)-WHITE-WHITE W/BLACK SPOTS-BLACK- POP.

BBLE WALL

$$\frac{1}{1,000,000} \text{ OF AN INCH}$$

Predicting the Pop

Now challenge the students to apply what they learned to invent a method for counting down, to the second, when their bubbles will pop. Here are strategies some students have used:

- timing how long a period elapses between the appearance of the first white color on the bubble and when it pops;
- noticing how far down the side of the bubble the transparent or "black" area extends before the bubble pops;
- noticing how long before the "pop" a bubble loses its reflective properties.

Let your students discover their own approaches before mentioning strategies used by other students.

Explaining the Phenomenon

The following explanations are written for the teacher. After your students have some success in predicting when a bubble will pop, you may want to discuss these explanations with them. Use your judgment about how much to present to your students. Typically, the concept of interference is first presented in high school physics courses.

1. Where do the colors in a bubble come from?

The colors in a bubble come from the reflection of white light shining on the bubble. White light contains waves of all different colors. The length of a wave, from crest to crest, determines its color. When light bounces off a bubble, some of each wave is reflected from the outer surface of the bubble wall, and some passes through to be reflected by the inner surface.

Interference refers to what happens when two waves pass through the same region of space at the same time. For example, when two rocks are thrown into a lake near each other, the two sets of circular waves *interfere* with one another. In some places, where the crest of one wave meets the crest of another, the motion of the water is increased. In other places, the crest of one wave meets the trough of another and there is little or no movement. The same basic process holds true for other wave motion, including sound waves and light waves.

When the thickness of the bubble wall is such that the two reflected parts of the wave of light leave the bubble in step, crest to crest (as illustrated by red light in the diagram), that color appears brighter *(constructive interference)*. Some colors of light will emerge crest to trough (as illustrated by blue light in the diagram) and will cancel each other *(destructive interference)*: those colors will not be seen. As the wall gets thinner, the colors that interfere constructively and destructively will also change.

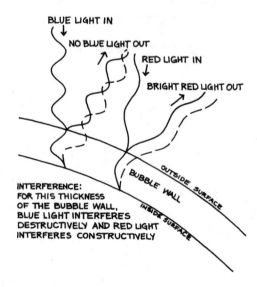

BLUE LIGHT IN

NO BLUE LIGHT OUT

RED LIGHT IN

BRIGHT RED LIGHT OUT

BUBBLE WALL

OUTSIDE SURFACE

INSIDE SURFACE

INTERFERENCE:
FOR THIS THICKNESS
OF THE BUBBLE WALL,
BLUE LIGHT INTERFERES
DESTRUCTIVELY AND RED LIGHT
INTERFERES CONSTRUCTIVELY

2. Why does the bubble appear white with growing black (or transparent) spots just before the bubble pops?

When the wall is less than a quarter wavelength thick for any color, none of the colors are completely cancelled—so the bubble appears white. Black spots appear when the wall is super-thin (about one millionth of an inch). This occurs because light reflected from the top surface is always reversed (all troughs become crests) but the light reflected from the back surface is not reversed. Thus, when the wall is super-thin, every light wave will cancel itself.

Cleanup

1. If dark table surfaces were used:

   a. First use a squeegie or paper towels to remove excess bubble solution from the table surface. **Do not add water.**
   b. Then sprinkle vinegar on the area to cut the soap film. Wipe dry with paper towels.
   c. Repeat once more if surface still retains soap film.

2. If trays with black construction paper were used:

   a. Discard soggy black paper.
   b. Pour the bubble solution remaining on the tray down the sink or into a spare container. (**Note:** black dye from the construction paper will have leached into the soap solution, causing it to appear dark in color.)
   c. Rinse off the tray.

Going Further

1. Assign students to look up the name Thomas Young and the concept of *interference*. They will find out about: (1) the controversy surrounding wave and particle theories of light; and (2) modern applications of interference phenomena, such as anti-reflection coatings on binoculars.

2. Have your students begin a collection of materials that exhibit the phenomenon of interference: abalone shells, peacock feathers, some sunglasses, etc.

Activity 6: Longer Lasting Bubbles

Introduction

Blow a bubble that will last for ten minutes... twenty minutes... or over an hour! Yes, it's possible! In 1917, Sir Thomas Dewar made a bubble that lasted 108 days. Since then, Eiffel Plasterer, a physicist who has been blowing bubbles for close to sixty years, blew a bubble that lasted for 340 days! Present this challenge to your students, and let them apply what they've learned about bubbles. This open-ended experiment serves as an excellent culmination to a unit on bubble science.

Note: Bubbles will often last longer than a single class period. If possible, plan on conducting this experiment in the morning, so students can check later in the day to see if their bubbles are still "alive."

What You Need

For preparation and cleanup:
- ☐ 8 oz. (240 ml) dishwashing liquid
- ☐ water
- ☐ 1 measuring cup or graduated cylinder
- ☐ 1 eyedropper
- ☐ 1 one-gallon container for mixing bubble solution
- ☐ glycerin (optional)

For each pair of students:
- ☐ 2 plastic drinking straws
- ☐ 1 pint-sized container (such as a cottage cheese container) for holding bubble solution
- ☐ 1 "Long-Lived Bubbles" data sheet (master included, page 44)
- ☐ 1 pencil

For the class:

Various materials such as:
- ☐ containers to blow bubbles in: clear screw-top jars—as large as possible, plastic dishpans, styrofoam egg cartons, etc.
- ☐ materials to cover containers: clear plexiglass sheets, plastic wrap, cafeteria trays, cheese cloth, aluminum foil
- ☐ humidifying materials: turkey basters, sponges, water spray bottles
- ☐ solution additives: sugar, glycerin, corn syrup, white glue, rubbing alcohol, extra dishwashing liquid
- ☐ volumetric measuring devices: measuring cups, measuring spoons, graduated cylinders, eyedroppers
- ☐ any other items you and your students deem appropriate

Getting Ready

1. Prepare one gallon of bubble solution:
   1 cup (240 ml) dishwashing liquid
   50–60 drops glycerin (optional)
   1 gallon water (3.8 liters)

2. Make one copy of the "Long-Lived Bubbles" data sheet for each pair of students.

3. Place all materials on a centrally located table.

Introducing the Challenge

1. Present your students with the problem: How long can you make a bubble last? Remind them that variables are things that can be changed. Ask them which variables might affect the life span of a bubble [humidity, air currents, solution additives]. Ask for their ideas of ways to change these variables in order to increase the life span of their bubble [increase the humidity of the air around the bubble, decrease a bubble's water loss by adding a hygroscopic substance to the bubble solution, keep a bubble in an enclosed container to reduce air currents, etc.].

2. Tell your students they will have two class sessions to work on the challenge. They can use any of the materials you've collected, and they may bring in additional materials from home.

Experimenting

Give each pair of students a data sheet. Tell them to use the data sheets to keep track of what they do and how long their bubbles last. Students should keep data sheets right next to their bubbles so ongoing observations can be recorded easily (e.g. 10:35—starting time. 10:45—bubble's alive. 10:55—alive. 11:55—alive, thinning on top, misted container with water, etc.). Because bubbles often outlive a single class period and occasionally last for several days, sometimes the teacher or other students will need to note whether a bubble is still "alive." Having the data sheet right there makes it easy for others to do this. School custodians have even been known to make data entries!

Reporting Results

Gather the group into a circle. Invite teams to report their results. Help the group to summarize.

- Which variables were most important in extending the life of a bubble?

- What strategies were most effective in altering those variables?

LONG-LIVED BUBBLES

NAME(S):_____

Describe the conditions of your experiment:	How long did the bubble last?

Resources

Boys, C.V. *Soap Bubbles; Their Colors and Forces Which Mold Them*, Dover Publications, 1959.
This classic bubble science text is for those who want to go right to "the source."

The Exploratorium, *Films, Foams, and Fizz*, 1983.
A very readable pamphlet containing explanations of bubbles and bubble phenomena.

Zubrowski, Bernie, *Bubbles*, Little Brown and Co., 1979. Though illustrated for elementary age children, this book is filled with good activities and questions for all grade levels.

Please see page 64 for a more extensive "bubble bibliography."

Summary Outlines

Activity 1: Bubble Technology

Getting Ready Before the Activity
1. Assemble materials.
2. Make bubble solution.
3. Cover tables with newspaper.
4. Fill containers with bubble solution and set on tables.
5. Place bubble-maker materials on tables.
6. Clear two table surfaces so students can classify materials.

Presenting the Challenge
1. Lead short discussion about blowing bubbles. Explain that students will discover which objects can be used to blow bubbles and what kinds of bubbles the objects make.
2. Point out objects they may test.
3. After testing each object they should place it on the "works" table or on the "doesn't work" table.
4. Write "small" on one side of the board and "large" on the opposite side. Have students write the name of each object that "works" on the board in order according to the bubble size produced.

Experimenting
1. Have students experiment with objects.
2. Circulate among students, posing questions.

Drawing Conclusions
1. Gather students around the chalkboard. Ask:
   - What's the same about all working bubble-makers?
   - What objects didn't work?
   - What were these objects lacking?
   - What strategies did you use to make something work?

2. Discuss which bubble-makers made large bubbles, which made small bubbles, and what characteristics those making larger bubbles have in common.
3. Explain that **technology** involves the use of science to create something practical. As a homework assignment, ask them to use what they've discovered about bubble-makers to create a specialized bubble-maker.

Activity 2: Comparing Bubble Solutions

Getting Ready Before the Activity
1. Assemble materials.
2. Duplicate one data sheet for each pair of students.
3. Label bubble solution containers with brand name.
4. Make three different bubble solutions (from three different brands of dishwashing liquid).
5. Fill labeled containers with bubble solution.
6. Clear one work station for each pair of students. Place a container of bubble solution and a meter stick at each station.

Introducing the Activity
1. Explain that students will compare brands of dishwashing liquids to find out which solution makes the biggest bubbles. Ask how they might do this.
2. Introduce the following procedure:
   a. Wet the table surface with bubble solution.
   b. Dip a straw into the bubble solution.
   c. Hold the straw near the soapy table surface and gently blow to form a bubble dome. Continue blowing until the bubble pops.
   d. Measure the diameter of the ring of soap suds left by the bubble dome.

Experimenting
1. Have students work in pairs to measure four bubbles per brand, record the bubble diameters on the data sheet, and calculate the average bubble diameter for each brand.
2. Respond to questions of "Is this fair?" by reminding them that they are the scientists and they must make sure that they treat each brand in exactly the same way.

Analyzing Results
1. Reconvene the class. Write the names of the brands on the chalkboard and have your students record their averages for each brand. Have the students calculate the grand average.
2. Make a bar chart of the results on the chalkboard. Rank the brands from biggest to smallest bubble-maker.
3. Ask if the test was a fair test. Introduce concept of **variable.** Explain concept of **controlled experiment.**
4. Ask and discuss why one brand might make bigger bubbles than another.

Activity 3: The Chemistry of Bigger Bubbles

Getting Ready Before the Activity
1. Assemble materials.
2. Duplicate one data sheet and one graphing sheet for each pair of students.
3. Make bubble solution.
4. Clear one work station for each pair of students.
5. Place demonstration materials on table or desk.
6. Make solutions "A" through "H."
7. Place all solutions and materials in centrally located area.

Observing Surface Tension
1. Lead a short discussion about substances that form bubbles.
2. Perform the following demonstration:
   a. Have the group squat around the demonstration table.
   b. Fill a glass to the top with water. Keep adding water drop by drop until the surface of the water is higher than the glass.

c. Tell students that they are observing the phenomenon of **surface tension,** and explain the concept. Tell students that surface tension prevents the water in the cup from spilling, and keeps bubbles from forming in many substances.

d. Demonstrate how a drop of bubble solution reduces surface tension. Explain that the surface tension in soapy water is low enough to allow bubbles to form easily.

Discussing the Problem of Evaporation

1. Explain that evaporation also affects the formation of bubbles.

2. Explain that substances with water-holding properties (**hygroscopic substances**) will reduce evaporation. Glycerin is a hygroscopic substance that slows the evaporation of water in bubble solution.

Planning the Experiment

1. Tell the class they will determine what effect the amount of glycerin in a bubble solution has on the size of the bubbles formed.

2. Help students plan a more controlled version of the experiment they used to compare brands of dishwashing liquid.

3. Describe the contents of Formulas "A" through "H" (1 cup bubble solution, 0 drops glycerin; 1 cup bubble solution, 10 drops glycerin, and so on). Ask students to predict which solution will make the biggest bubbles.

Experimenting

1. Assign pairs of students to test the solutions, using the same method they used in Activity 2, with whatever improvements the class decided to make.

2. Have students swap work stations and test additional formulas until each formula has been tested by at least 4 student teams.

Graphing the Results

1. Record the averages on the board. Calculate the grand average for each formula.

2. Ask your students to graph the results of all experiments on the graphing sheet. What is the optimum amount of glycerin for making the biggest bubbles?

3. Ask if anyone was surprised by the results.

Activity 4: Bernoulli's Bubbles

Getting Ready Before the Activity
1. Assemble materials.
2. Make bubble solution.
3. Fill containers with bubble solution and place them on squares of newspaper in various locations.
4. Spend five minutes practicing the technique of keeping a bubble aloft by waving your hand back and forth over it.

Exploring
1. Challenge students to devise ways to keep a bubble from hitting the ground without touching the bubble.
2. Have students set their bubble-blowers aside. Make two lists: methods that work, and those that do not.

Explaining
1. Provide a brief historical perspective on human pursuit of flight, and the Bernoulli principle. Explain that the Bernoulli principle states that the faster air flows, the less pressure it exerts.
2. Draw a diagram of an airplane and an airplane wing on the board and explain how the Bernoulli principle applies to airplanes.
3. Tell students that the force pushing upward is called **dynamic lift.**
4. Explain that there are two ways to keep an object aloft: increase the pressure under it, or decrease the pressure over it. Have the class examine the list of methods that kept a bubble aloft, and discuss how each method worked.

Applying the Principle
1. Demonstrate how to use Bernoulli's principle to decrease the pressure over a bubble.
2. Challenge your students to use Bernoulli's principle to keep their bubbles aloft.
3. Distribute index cards for students to use in waving over their bubbles.

Activity 5: Predict-A-Pop

Getting Ready Before the Activity
1. Assemble materials.
2. Make bubble solution.
3. Fill containers with bubble solution and place in central location with straws, white paper, and masking tape.
4. Clear a work surface for each group of 2–4 students. If the surface is not dark, line a cafeteria tray with black construction paper.
5. Prepare one "white collar."

Observing Colors
1. Demonstrate the procedure:
    a. Pour bubble solution on the dark surface or on the tray lined with construction paper.
    b. Place the white collar around the soapy area.
    c. Dip a straw into the soap solution.
    d. Hold the straw near the soapy table surface and gently blow to form a bubble dome.
    e. Remove the straw.
2. Instruct each group of 2–4 students to make a collar, blow a bubble dome, and observe the changing colors on top of the bubble. Tell them to record the sequence of colors they see for 4 or 5 bubbles.

Reporting Results
1. Assemble the students near the chalkboard. Record the findings of several teams on the board. Explain that the colors on the surface of a bubble change as the bubble becomes thinner and thinner.
2. A typical sequence that scientists have noticed is: green—blue—magenta—yellow—(sequence repeats more than once)—and finally white— white with black spots—black—POP! Explain that not all students will see this pattern because air currents interfere with the thinning of the bubble.
3. Draw a diagram of the decreasing bubble wall under a listing of the typical color sequence. Ask students if they notice any aspects of the typical pattern in their data.

Predicting the Pop
Challenge students to apply what they have learned to invent a method for counting down, to the second, when their bubbles will pop.

Explaining the Phenomenon (Optional)
1. Where do the colors in a bubble come from?
   a. The colors come from the reflection of white light.
   b. White light contains waves of all different colors.
   c. The length of a wave, from crest to crest, determines its color.
   d. When light bounces off a bubble, some of each wave is reflected from the outer surface of the bubble wall, and some passes through to be reflected by the inner surface.
   e. **Interference** refers to what happens when two waves pass through the same region of space at the same time.
   f. When the thickness is such that reflected waves leave crest to crest, that color appears brighter (**constructive interference**).
   g. When reflected waves leave crest to trough, that color will cancel (**destructive interference**).
   h. As the bubble wall gets thinner and thinner, the colors that interfere constructively and destructively also change.
2. Why does the bubble appear white with growing black spots just before the bubble pops?
   a. When the wall is less than a quarter wavelength thick for any color, none of the colors are cancelled completely, so the bubble appears white.
   b. Black spots appear when all colors are cancelled.

Activity 6: Longer Lasting Bubbles

Getting Ready Before the Activity
1. Assemble the materials.
2. Make the bubble solution.
3. Duplicate one data sheet for each pair of students.
4. Place all materials on a centrally located table.

Introducing the Challenge
1. How long can you make a bubble last?
   a. Which variables might affect the life span of a bubble?
   b. How could you change those variables to increase the life span of a bubble?
2. Announce that students have two sessions to work on this challenge, and may want to bring additional materials from home.

Experimenting
1. Distribute a data sheet to each pair of students.
2. Explain how to use the sheet to record results.

Reporting Results
1. Invite teams to report their results.
2. Help the group summarize which variables were most important in extending the life of a bubble and what strategies were most effective in altering those variables.

Assessment Suggestions

Selected Student Outcomes

1. Students discover the properties of soap film and the ways that soap film creates bubbles and then apply this knowledge in a variety of situations.

2. Students learn to distinguish fair and unfair elements of a test.

3. Students improve their ability to design controlled experiments.

4. Students explain how evaporation and surface tension impact a liquid's ability to make bubbles.

5. Students predict how air currents affect bubbles in a variety of situations.

6. Students apply their knowledge of color patterns to predict when a bubble will pop.

Built-In Assessment Activities

Design a Bubble Maker: As a homework assignment at the end of Activity 1, Bubble Technology, students design and draw bubble makers for specialized uses. The teacher can see how students apply their knowledge of soap films as they describe their bubble makers to the class. (Outcome 1)

Designing Other "Fair Tests": In Activity 3, The Chemistry of Bigger Bubbles, students are asked how they would design the experiment so it is a fair comparison. The teacher can see how students identify uncontrolled variables, and how they devise a plan to control those variables. (Outcomes 2, 3)

Predicting the Pop: In Activity 5, Predict-A-Pop, students are challenged to apply what they've learned to invent a method for counting down, to the second, when their bubbles will pop. As students repeat this procedure, the teacher can observe how students use finer and finer observations to modify their predictions. (Outcome 6)

Longer-Lasting Bubbles: In Activity 6, Longer-Lasting Bubbles, students are challenged to apply what they've learned to blow a bubble that will last as long as possible. The teacher can see what factors students consider as they design systems to maintain bubbles. Teachers can also observe whether students use fair or unfair tests to approach this challenge. (Outcomes 1–6)

Assessment Suggestions (continued)

Additional Assessment Ideas

Dishwashing Tests: In the Going Further suggestion for Activity 2, Comparing Bubble Solutions, students design an experiment that tests the dishwashing properties of various brands of dishwashing liquid. The teacher has the opportunity to observe how students identify variables and plan an experiment that has controls for those variables. (Outcomes 2, 3)

The Ideal Bubble Solution: Students can make recommendations to the product development specialist at a toy company that plans to produce bubble toys. Their advice needs to include discussion of the following two points. (1) Why water needs soap in order to make bubbles that last. (2) Why bubble solution with glycerin makes sturdier bubbles. (Outcome 4)

Sharing Bubble Discoveries: Students can write a letter to someone who has never blown soap bubbles before. They can include five different things they learned about bubbles such as tips for blowing table bubbles, how to blow the biggest bubble, how to keep a bubble in the air, how to make a bubble last long, or anything else that will let the person know about the properties of bubbles. (Outcomes 1, 4, 5, 6)

Bubble Maker Video Commercial: If your class has access to a video camera, have students write, design, narrate, and produce a short commercial to advertise their originally designed bubble maker. The commercial should explain to the viewer why the bubble maker makes the kinds of bubbles it does. (Outcome 1)

Literature Connections

Splash! All About Baths by Susan K. Buxbaum and Rita G. Gelman. Illustrated by Maryann Cocca-Leffler. Little, Brown and Co., Boston. 1987. Grades: K–6. Before he bathes, Penguin answers his animal friends' questions about baths such as, "What shape is water?" "Why do soap and water make you clean?" "What is a bubble?" "Why does the water go up when you get in?" "Why do some things float and others sink?" and other questions. Answers to questions are clear. The book received the American Institute of Physics Science Writing Award.

Glorious Flight by Alice and Martin Provensen. The Viking Press, New York. 1983. Grades: 2–4. The story of Louis Bleriot, an aviation pioneer who developed and flew his aircraft over the English Channel in 1909. Drawings show the evolution of the various prototypes of flying machines. Relates to "Activity 4: Bernoulli's Bubbles" in which students gain direct experience with Bernoulli's principle.

Danny Dunn and the Heat Ray by Jay Williams and Raymond Abrashkin. Illustrated by Owen Kampen. McGraw-Hill, New York. 1962. Grades: 4–6. Danny and his sidekick Irene Miller explore various science fair project possibilities. They choose one that demonstrates how airplanes fly. The story includes an explanation of dynamic lift that relates perfectly to "Activity 4: Bernoulli's Bubbles."

The Toothpaste Millionaire by Jean Merrill. Illustrated by Jan Palmer. Houghton Mifflin Co., Boston. 1972. Grades: 5–8. Twelve-year-old Rufus Mayflower doesn't start out to become a millionaire, as incensed by the price of a tube of toothpaste, he tries making his own from bicarbonate of soda with peppermint or vanilla flavoring. Assisted by his friend Kate (who narrates the story) and his math class (which becomes known as Toothpaste 1), his company grows from a laundry room operation to a corporation with stock and bank loans. In Activity 1 of "Bubble-ology" students design specialized bubble-makers. Beginning on page 47, Rufus designs a machine for filling toothpaste tubes.

Chitty Chitty Bang Bang: The Magical Car by Ian Fleming. Illustrated by John Burningham. Random House, New York. 1964. Grades: 6–adult. Series of adventures featuring a transforming car, an eccentric explorer and inventor, and 8-year-old twins. Nice combination of technical and scientific information, much of it accurate, with a more mystical sense of how some machines seem to have "a mind of their own." Relates to Activity 1: Bubble Technology in which students are challenged to invent specialized bubble makers.

Hailstones and Halibut Bones: Adventures in Color by Mary O'Neill. Illustrated by John Wallner. Doubleday, New York. 1961. Grades: All. Twelve two-page poems provide the author's impression of various colors. Her perceptions go far beyond visual descriptions, painting a full spectrum of images. Connects nicely to "Activity 5: Predict A Pop" in which students observe the color changes that take place before a bubble pops.

Bubbly Background and "Bubbliography"

Bubbles and their behavior have fascinated people of all ages from time immemorial. Whether it be in the foam upon the sea, the froth of root beer, or the first reflection of a soaring soap bubble in a baby's bright and curious eyes, bubbles are the "stuff" of poetry, of wonder and whimsy.

Yet bubbles are equally the "stuff" of science and mathematics. Their behavior has attracted and compelled observation by chemists, physicists, mathematicians, and engineers. The search for deeper understandings about bubbles has yielded intriguing connections in many technical fields. Many books and articles have been written to consider these connections.

The bibliography at the end of this section can provide greater detail on specific subjects. This background is not meant to be read out loud to your students; it is provided to help you consider student questions and to give you a glimpse into this rich and diverse subject, as we skim along the shining surface to look into the scientific and mathematical marvels and forces that dance within and are reflected from—the bubble.

What is a Bubble?

A bubble is a thin skin of liquid surrounding a gas. This thin skin, or, in the case of soap bubbles, this *soap film*, has elastic qualities; it can stretch. The soap film is composed of molecules of water and soap. In the case of soap bubbles, the "gas" that the soap film surrounds is composed of either the gases that make up the air, or, in the case of bubbles we blow ourselves, the carbon dioxide and other gases that we exhale. The liquid and gas are separated by a *surface*, the soap film.

Surface Tension

What gives the combination of soap and water its unique, bubble-producing qualities? *Surface tension* is an important factor. The surface of water is always in a state of strong tension. We are all familiar with common manifestations of water's surface tension, such as when we see a water bug able to glide across the surface of the water. Or perhaps you've presented experiments to your class in which water swells out over the edge of a penny, or a paperclip floats. These phenomena take place because of surface tension.

If you watch a drop slowly forming on the end of a leaky water faucet you can see how it gradually changes shape, with the taut surface of the water drop at first containing the emerging water, then it begins to bulge out, then finally, when the surface tension of the water can no longer resist the downward pressure, a round drop detaches and falls. The forces of tension in the surface of the water at first contained the drop, but when it fell, the surface tension all around the drop then forced it into a sphere. (Another liquid, with less surface tension, would behave differently. For example, in Session 3 of the GEMS teacher's guide *Liquid Explorations*, students compare drops of oil and water, discovering that while water drops are round, oil drops are flat! Surface tension accounts for this difference.)

Surface tension is caused by the attraction a substance has to itself. Some molecules, such as water, have a positive electrical charge on one end and a negative charge on the other. These molecules can align, particularly on the surface of a liquid, so the positively charged end of one molecule forms an "attachment" with the negatively charged end of another molecule. These attachments are known as *"weak bonds."* Their existence is what gives water its "stickiness" and what accounts for surface tension, because at the surface the bonds with other water molecules make the water behave as if it had a stretchy skin.

Along Comes Soap

When *soap* enters the scene, the two-sided nature of its molecules (with one side that is attracted to water and another side that is repelled by water and attracted to grease) creates important changes by *lessening* water's surface tension. In fact, soap reduces water's surface tension to about 1/3 of what it usually is and some detergents cut the surface tension more than that. Here's how the reduction of surface tension happens:

(1) Soap molecules and water molecules are crowded together at the water's surface. (2) The soap molecules near the surface have the water-attracting end next to the water molecules, but their water-repelling ends seek to push outward, into the air, as they are repelled by the water molecules. (3) As these water-repelling ends of the soap molecules push outward, they also push between water molecules, spreading them apart. The wider apart these surface water molecules are, the less strong the hydrogen bonds between them. Thus, soap weakens water's surface tension.

The prevalence of the water-repelling (grease attracting) ends of the soap molecules near the surface also helps prevent the water from evaporating. This slowing down of *evaporation* is another factor that contributes to the ability of the soap film, the mixture of soap and water, to produce bubbles that last for some time.

Many bubble solutions also call for the addition of glycerin. Glycerin is a hygroscopic substance, meaning that it holds water, slowing evaporation. It thus intensifies the evaporation-slowing properties of soap. There is, however, an optimum amount, because *too much* glycerin can begin to have a counterproductive effect, and not contribute to making better bubbles. In Activity 3 of *Bubble-ology*, students experiment with differing amounts of glycerin to determine the optimum amount for making lasting bubbles.

Some people think that soap in combination with water helps make bubbles because it *increases* water's surface tension, but, as we have described, **the opposite is true.** In fact, two reasons why lasting bubbles cannot be made from water alone is that the surface tension of water is too strong and because evaporation takes place too fast. Soap and water together are loose and elastic enough to be stretched into a bubble, while water alone is not. **Soap reduces the water's surface tension and slows its evaporation, making it an ideal bubble-producer when mixed with water.**

Soap films also have the property of being self-healing. As long as a finger or other object that is brought into contact with the soap film is also wet with soap solution, a moderately thick soap film will flow onto the wet finger or other object, thereby keeping the soap film continuous, and not popping the bubble.

Shapes and Surfaces

A bubble's **shape** is governed in part by the force of surface tension. The soap film enclosing a certain quantity of air stretches **only as far as it must** to balance the air pressure inside the bubble against the surface tension of the soap film. The bubble thus represents a *state of equilibrium* between air pressure and surface tension. The result is the spherical shape of bubbles. **The liquid forms the shape that encloses the greatest volume within the least surface area. This shape is a** *sphere*.

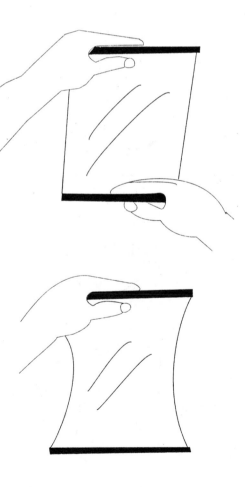

Soap films also provide excellent examples of the mathematical and physical concept of minimal surfaces. If you gently blow at or twist the surface of a film stretched across a wire frame, it will stretch and its area will increase, but when you stop pushing, the soap film springs back to its original shape, taking up the smallest possible area that it can while still spanning the frame it is in. Pulling on the "handles" of a "bubble window" can make a true square or rectangle, but when left to hang, the soap film pulls the strings inward, resulting in a lower energy situation and a minimal surface arrangement (see illustration). Thinking about bubbles and other phenomena that behave in this way introduces topics like measurement of surface area, the relation of surface to area (which is always proportional in a bubble) and other physical science, mathematical, and geometrical concepts.

The edge of a soap film stretched across a wire frame is actually a solution to a complicated mathematical problem—finding the shortest possible path while obeying the laws of gravity and attaching to all surfaces dipped in soap. A very interesting example of this is provided when a wire frame in the shape of a cube is dipped in soapy water. The soap film sides that form can be forced to collapse in on the center, becoming a bulging cube that is untouched by the frame itself. The bubble tries to be round, but it cannot free itself from the frame. This has been described by Eiffel Plasterer as "a beautiful geometric compromise between a cube and a sphere." (Plasterer is perhaps the most famous "bubble-ologist" in the world; he holds the record for longest-living bubble—340 days—and has experimented with all kinds of bubble solutions and materials.) Others have compared a simple soap bubble to a computer, in the sense that soap film and bubbles instantly solve complex mathemetical problems involving minimum paths and mimimum surfaces with maximum volume, problems that would require a powerful computer to solve, and then not instantly. This is one of the reasons why so many scientific and technical disciplines have at one time or another used bubbles as models and simulations for physical or biological processes.

Polygons and Polyhedrons

In the terms of modern physics, the soap film's capacity to seek minimal surfaces provides an excellent demonstration of a **system's** tendency to seek the state of lowest energy at which it can still function as a system. Soap bubbles are a system of interacting parts that include the liquid (soap and water), the gases enclosed, gravity, air pressure, surface tension, other properties, etc. This minimizing of both surface area and systemic energy while maximizing volume also affects the ways that soap bubbles cluster together. When two bubbles meet, a smooth sheet separates them that is flat if the bubbles are the same size. When three bubbles come together, three surfaces intersect to form a line, and the angle between each pair of sheets is always 120°. Exploration of angles and their intersection, a key geometric concept, is another rich bubble math extension. The hexagonal-shaped bubbles formed by the intersection of bunched bubbles in the bubble wall are caused by this convergence of three sides. Interestingly, the hexagon is an efficient structure in nature, notably seen in the design of honeycomb.

But it doesn't stop there! The clustering of these hexagonal shapes, the stacking of bubbles, produces polyhedrons, or three-dimensional polygons. The German astronomer Johannes Kepler visualized the hexagonal crystalline pattern that comes from subjecting spheres to pressure, and he conceived of a twelve-sided figure with diamond-shaped faces like crystal garnet, called a rhombic dodecadron. Compressed eggs and cells tend toward this shape; combining maximum volume with less partitional area than almost any other multi-sided shape—nature at its most efficient.

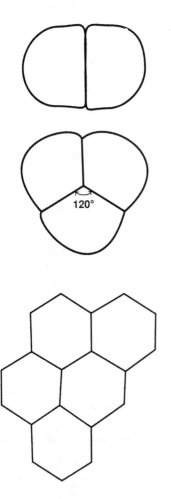

120°

Lord Kelvin, the Scottish mathematician, then proposed a 14-sided figure, a tetrakaidekahedron (TKH) as even more ideal for living cells, with a slightly smaller percentage of partitional surface. It took mathematicians a long time to conceptualize these "many-sided" ideas, but bubbles, cells, and crystals were already doing it! So, even in advanced realms of upper level mathematics, crystalline formation, and cell biology, scientists have learned and continue to explore a lot from the behavior of bubbles!

Topology and Architecture

The special properties of soap film, shown in numerous experiments with intricate loops and shapes and soap film, can introduce the fascinating subject of *topology*, which is the exploration of the geometrical properties of various shapes, substances, and figures, especially relating to what happens when shapes, substances, and figures are bent, stretched, or molded.

Using soap films and soap-film computer simulations, mathematicians have found complex minimal surface forms called catenoids and helicoids, and their work has already found adaptations in dentistry, embryology, and architecture. For example, architect Frei Otto, who designed structures at the Munich Olympics, utilized soap film models to design graceful, airy, and web-like structures, using as little construction material as safely possible to create exhibition halls, arenas, and stadiums that could be easily built, dismantled, and moved. Otto has also written a book on tensile structures that we list below.

Bernoulli's Principle and Aerodynamics

In Activity 4 students are challenged to keep a bubble aloft and in so doing gain hands-on understanding of an important scientific principle named for Daniel Bernoulli, an 18th century Swiss scientist, who published a statement of the principle in 1738. Although Bernoulli's work concerned liquids, the principle he articulated can be applied in many ways, including to the field of aerodynamics. It is an early statement of the general law in physics of the conservation of energy.

As regards a gas, or a mixture of gases, such as air, the Bernoulli principle can be briefly expressed as: "the faster air flows, the less pressure it exerts." The questions for students and the drawing on page 31 of this guide provide an explanation of how this principle leads to the *dynamic lift* that plays an important role in making airborne flight possible. In accordance with Bernoulli's principle, the faster moving air that travels over the top of the wing exerts less pressure than the air that travels beneath the wing, thus creating an upward push.

Light and Color

The beautiful color patterns that are seen when a beam of light shines through a soap bubble are produced whenever light passes through extremely thin layers. This is similar to an oil slick on the street, or even a peacock's tail feathers. The colors come from the reflection of white light shining on the bubble. White light contains waves of different colors—the spectrum. The length of a wave determines its color. When light bounces off the bubble, some of each wave is reflected from the outer surface of the bubble wall, and some passes through the bubble and is reflected by the inner surface. This means there are two sets of light waves passing through the same space at the same time, and causes a phenomenon called *interference*. Sometimes the crests of the two sets of waves coincide, intensifying both (*constructive interference*); sometimes the crest of one meets the trough of another and that particular color is cancelled out (*destructive interference*). As the bubble wall thins, the wavelengths that are seen or not seen also change.

Just before it bursts, the bubble appears white with a growing black (or transparent) spot. This is because, when the wall is less than a quarter wavelength thick, none of the colors is completely cancelled out, all are present, so the bubble appears white. A black spot or dot appears when the bubble is super-thin (about one-millionth of an inch) because, due to the (destructive) interference phenomenon at that point, all the waves from the front surface and back surface cancel each other out. The black dot gradually expands, then disappears as the bubble pops.

There are many experiments that can be done exploring the fascinating and colorful interactions between light and color. Understanding the interaction of light, color, and materials has many industrial applications. In *Bubble-ology* , students "Predict-a-Pop" in Activity 5, gaining practical experience and getting a great introduction to important concepts in physics while exploring the same wondrous phenomena represented by the rainbow.

A Vast Spectrum of Real-Life Applications

In the standard tool chest, we find many types of levels that utilize bubbles. You could bring one or more of these into class. Bubbles are also in some sextants, used in navigation and surveying.

As mentioned above, bubble shapes, minimal energy and surface qualities, the formation of polyhedrons, Bernoulli's principle, and other bubble behavior has found application in many engineering, aerodynamic, and architectural fields, including the construction of tents and geodesic domes, larger structures, and as models for the solution of a wide range of engineering problems. A good example is designing a tin soup can—what is the best shape and minimum amount of material that can be used in packaging? Bubble have helped engineers figure this out.

Even bubble noise, the sound of tiny air bubbles bursting, has been explored as a way of detecting submarines and is known by scuba divers as a high-pitched tinkling sound. Liquid drop models, one form of bubbles, have been used in mathematical models studying the nucleus of the atom.

The study of bubble formation and behavior, including "soap froth," has also provided useful models for further understanding the shape of developing embroyos, the chemical structure of polymers, physical processes in crystallography, and the structure of metals. In dentistry, it has been suggested that least-surface area shapes could be used to design bone implants for securing false teeth, to minimize contact with the bone while maximizing a strong bond.

We are all familiar with the rushing bubbles in boiling water. In nuclear physics, scientists use the phenomena of boiling bubbles in a bubble chamber to track the pathways in invisible atomic particles. The paths left by high-speed protons going through liquid hydrogen as the pressure is reduced just enough for the very first boiling to occur are tiny trails of bubbles. The invention of the bubble chamber by Donald Glaser (who received the Nobel Prize in 1955) has helped make possible many new developments in particle physics and further widened our understanding of atomic structure.

But there is no need to look far afield for bubbles. From the bubbles in carbonated drinks, beer, or sparkling champagne, to the yeast bubbles that form in the baking of bread, bubbles are part of our everyday life. Hot air balloons and blimps were the first ways people invented to fly. Air bubbling into an aquarium partly dissolves into the water and enables the fish to breathe. A lightbulb is a bubble of inert gas that keeps oxygen away from a glowing filament, preventing it from burning up quickly. Firemen use foam to put out fires, blocking off oxygen, and releasing cooling moisture as the bubbles pop. Plastic bubble "pak" is used for mailing fragile items (and is fun to play with besides!).

Last but perhaps not least, it has been suggested by several scientists, including an astrophysicist named Margaret Geller, that some astronomical computer tracking and computation indicates that the entire Universe itself may be composed of bubbles. These theories speculate that the planets, stars, and galaxies we know are actually on the surface of these giant bubbles. It is even possible that these bubbles may behave like soap bubbles, sometimes merging into each other!

More About Bubbles: A Bubble Bibliography

Almgran, Frederick J. Jr., and Taylor, Jean E. "The Geometry of Soap Films and Soap Bubbles." *Scientific American.* (July 1976) 235: 82.

"Bubbles, Foam, and Fizz." *Ideas in Science.* (January 1983).

"Geometry in Nature: Bubbles/Mathematics in Nature." *Ideas in Science.* (Vol. 3, No. 2, 1986).

Bohren, Craig F. *Clouds in a Glass of Beer: Simple Experiments in Atmospheric Physics.* New York: John Wiley & Sons, 1987.

Boys, C.V. *Soap Bubbles: Their Colors and Forces Which Mold Them.* New York: Dover, 1959.

Chemistry with Bubbles. Mountain View, California: The New Curiosity Shop, Inc., 1988. (In cooperation with Apple Computer, Inc.).

Clift, R. Grace, J.R., and Weber M.E. *Bubbles, Drops, and Particles.* San Diego: Academic Press, 1978.

"Bubbles." The Exploratorium Magazine. (Winter 1982). (Copies available from The Exploratorium, 3601 Lyon Street, San Francisco, CA 94123.)

Isenberg, Cyril. *The Science of Soap Films and Soap Bubbles.* New York: Dover, 1978.

Katz, David A. *Chemistry in the Toy Store.* Department of Chemistry, Community College of Philadelphia. Second edition, 1983. (Copies available from the author.)

Levine, S., Strauss, M.J., Mortier, S. "Soap Bubbles and Logic." *Science and Children.* (May 1986) 23:10-12.

Murphy, Jamie. "Bubbles in the Universe." *Time.* (January 20, 1986): 51.

Otto, Frei. *Tensile Structures.* Cambridge, Massachusetts: Massachusetts Institute of Technology Press, 1972.

Peterson, Ivars. "Science Meets the Soap Bubble." *Washington Post.* (September 18, 1988): C3.

Rice, K. "Soap Films and Bubbles." *Ideas in Science.* (May 1986): 23:4-9.

Siddeons, Colin. "Soap Bubble Spectra." *The Science Teacher.* (January 1984): 26.

Steinhaus. H. *Mathematical Snapshots.* New York: Oxford University Press, 1983.

Stevens. Peter S. *Patterns in Nature.* Boston: Little, Brown and Company, 1974.

Zubrowski, Bernie. Illustrated by Joan Drescher. *Bubbles: A Children's Museum Activity Book.* Boston: Little, Brown and Company, 1979.

Zubrowski, Bernie. "Memoirs of a Bubble Blower," *Technology Review.* (November/December 1982).

BUBBLE SOLUTIONS

NAME(S):_____

Soap Brand_____

Bubble 1_____ 2_____

3_____ 4_____

Average Bubble Diameter_____

Soap Brand_____

Bubble 1_____ 2_____

3_____ 4_____

Average Bubble Diameter_____

Soap Brand_____

Bubble 1_____ 2_____

3_____ 4_____

Average Bubble Diameter_____

WHICH SOAP BRAND HAD THE BIGGEST AVERAGE BUBBLE DIAMETER?

EXPERIMENTING WITH GLYCERIN

NAME(S): _____

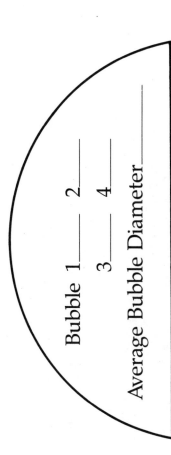

Bubble 1 _____ 2 _____

3 _____ 4 _____

Average Bubble Diameter _____

Number of Drops of Glycerin _____

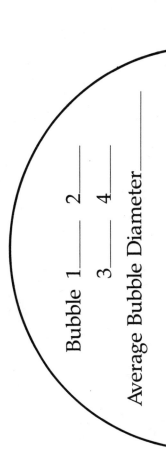

Bubble 1 _____ 2 _____

3 _____ 4 _____

Average Bubble Diameter _____

Number of Drops of Glycerin _____

Bubble 1 _____ 2 _____

3 _____ 4 _____

Average Bubble Diameter _____

Number of Drops of Glycerin _____

Bubble 1 _____ 2 _____

3 _____ 4 _____

Average Bubble Diameter _____

Number of Drops of Glycerin _____

GRAPHING SHEET

NAME(S)_____

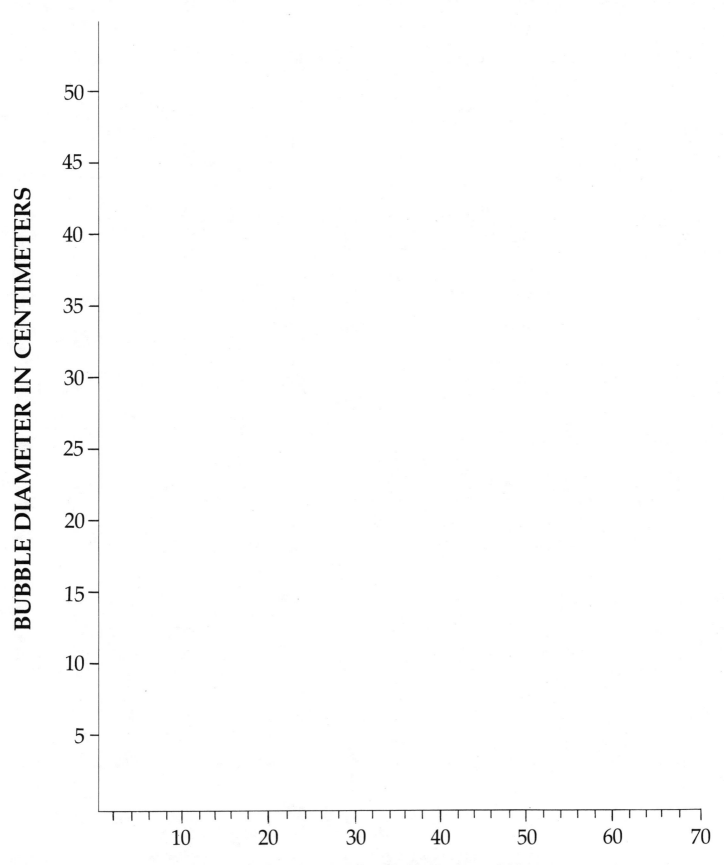

BUBBLE DIAMETER IN CENTIMETERS

NUMBER OF GLYCERIN DROPS

LONG-LIVED BUBBLES

Describe the conditions of your experiment:	How long did the bubble last?

Notes